JUMBO COLORING BOOK

MARINE ECOSYSTEMS ARE AQUATIC ENVIRONMENTS WITH HIGH LEVELS OF DISSOLVED SALT.

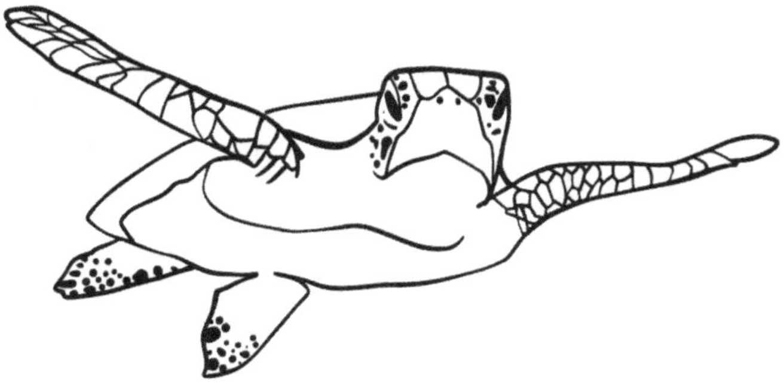

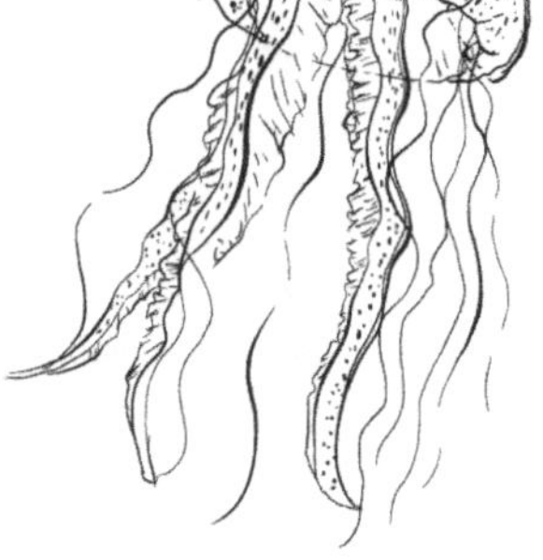

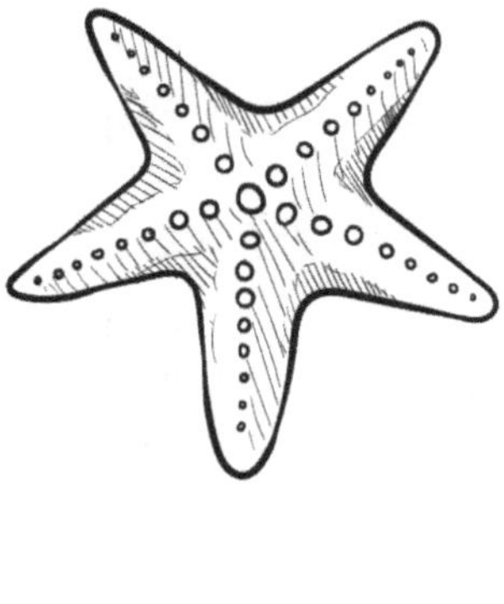

Biographical Note
Marine Ecosystems An Educational Coloring Book is a new work,
first published by Little Artist Studio in 2025.

International Standard Book Number
ISBN 979-8-9992504-0-7

www.littleartiststudio.org

DIVE INTO THE WONDERS OF THE WORLD'S MARINE
ECOSYSTEMS WITH OVER 85 STUNNINGLY DETAILED
COLORING PAGES. THIS EDUCATIONAL COLORING
BOOK FEATURES ICONIC SCENES FROM OCEANS
AROUND THE GLOBE, DESIGNED TO INSPIRE
CURIOSITY AND DEEPEN UNDERSTANDING OF OUR
PLANET'S VITAL MARINE ENVIRONMENTS. AS PART OF
LITTLE ARTIST STUDIO'S CELEBRATED EDUCATIONAL
SERIES, EACH FULL-PAGE ILLUSTRATION TELLS A
CAPTIVATING STORY THAT FUELS CREATIVITY AND
LEARNING. PRINTED ON SINGLE-SIDED PAGES, THIS
COLORING BOOK ALLOWS ARTISTS OF ALL AGES TO
EXPERIMENT WITH ANY MEDIUM AND EASILY
SHOWCASE THEIR VIBRANT CREATIONS. IDEAL FOR
NATURE ENTHUSIASTS, EDUCATORS, AND CREATIVE
EXPLORERS ALIKE.

WETLAND

A WETLAND IS A PLACE WHERE THE
LAND IS COVERED WITH WATER,
EITHER ALL THE TIME OR JUST
PART OF THE YEAR.

WETLAND

THERE ARE DIFFERENT TYPES OF
WETLANDS, AND ONE WAY TO
GROUP THEM IS BY WHERE THEY
ARE FOUND.

MARINE WETLAND

A MARINE WETLAND IS A PLACE NEAR THE OCEAN WHERE LAND IS OFTEN WET, AND THE WATER IS SALTY OR A LITTLE SALTY. MANY PLANTS AND ANIMALS LIVE THERE.

MARINE WETLAND

MARINE WETLAND

SOME FISH, LIKE BABY SHARKS AND TINY CLOWNFISH, GROW UP IN MARINE WETLANDS BECAUSE THEY'RE SAFE HIDING SPOTS FROM BIGGER PREDATORS!

MARINE WETLAND

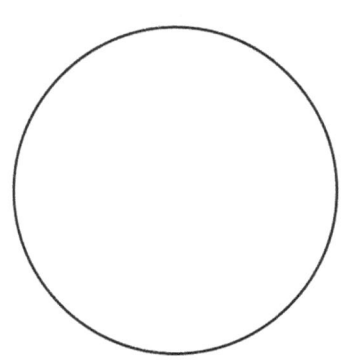

MARINE WETLANDS HELP CLEAN
DIRTY WATER BY TRAPPING
POLLUTION IN THEIR PLANTS
AND MUD.

MARINE WETLAND

THE GRAND CANAL CONNECTS VENICE TO THE
VENETIAN LAGOON, A MARINE WETLAND WHERE
SALTY SEA WATER AND FRESH RIVER WATER MIX.
IT'S A SHALLOW, WATERY AREA FULL OF PLANTS
AND ANIMALS THAT ALSO HELPS PROTECT THE
CITY FROM FLOODS.

ESTUARINE WETLAND

AN ESTUARINE WETLAND IS A SPECIAL KIND OF WETLAND FOUND WHERE A RIVER MEETS THE OCEAN. THIS PLACE IS CALLED AN ESTUARY.

ESTUARINE WETLAND

MALLARD

ESTUARINE WETLAND

THE WATER IN AN ESTUARY IS A MIX OF FRESH WATER (FROM THE RIVER) AND SALT WATER (FROM THE SEA).

ESTUARINE WETLAND

ESTUARINE WETLAND

ESTUARINE WETLANDS ARE FULL
OF MUDDY OR GRASSY AREAS THAT
GET COVERED WITH WATER WHEN
THE TIDE COMES IN AND
UNCOVERED WHEN IT GOES OUT.

RIVER OTTER

ESTUARINE WETLAND

RIVERINE WETLAND

A RIVERINE WETLAND IS A WETLAND THAT FORMS NEXT TO A RIVER. IT GETS WET WHEN THE RIVER GETS HIGHER OR FLOODS.

RIVERINE WETLAND

RIVERINE WETLAND ARE FOUND ALONG THE SIDES OF RIVERS AND ARE GREAT PLACES FOR ANIMALS TO LIVE.

BEAVER

RIVERINE WETLAND

RIVERINE WETLAND

A WATERFALL ISN'T A WETLAND, BUT
RIVERINE WETLANDS CAN FORM JUST
BELOW IT, WHERE THE WATER SLOWS
DOWN AND CREATES A PERFECT HOME
FOR PLANTS, FROGS, AND FISH!

RIVERINE WETLAND

GREAT BLUE HERON

LACUSTRINE WETLAND

LACUSTRINE WETLANDS ARE WET, SHALLOW AREAS NEXT TO LAKES WHERE THE WATER IS STILL AND NOT FLOWING

LACUSTRINE WETLAND

LACUSTRINE WETLAND

LACUSTRINE WETLANDS ARE LIKE
NATURE'S SPONGE-THEY SOAK UP EXTRA
WATER FROM RAIN AND HELP KEEP LAKES
CLEAN BY FILTERING OUT POLLUTION.

PALUSTRINE WETLAND

A PALUSTRINE WETLAND IS A WETLAND
THAT IS NOT NEAR THE OCEAN, RIVER,
OR LAKE. IT'S USUALLY FOUND INLAND
AND LOOKS LIKE A MARSH, SWAMP, OR
BOG.

PALUSTRINE WETLAND

A PALUSTRINE WETLAND IS A WET
AREA WITH SHALLOW WATER, PLANTS
LIKE GRASSES OR TREES, AND IT CAN
BE WET ALL THE TIME OR JUST WHEN
IT RAINS.

PALUSTRINE WETLAND

A PALUSTRINE WETLAND IS
IMPORTANT BECAUSE IT GIVES
ANIMALS A HOME AND HELPS STORE
RAINWATER AND KEEP THE
ENVIRONMENT CLEAN.

BALD CYPRESS

TIDAL WETLANDS

TIDAL WETLANDS ARE SPECIAL
BECAUSE THEY HAVE SALTWATER
AND GET FLOODED BY OCEAN TIDES.

TIDAL WETLANDS

TIDAL WETLANDS ARE
CLASSIFIED BY THE AMOUNT OF
WATER COVERING THE AREA AT
HIGH AND LOW TIDES AND THE
TYPE OF VEGETATION.

TIDAL WETLANDS

GREAT BLUE HERONS
NEST IN COLONIES IN
TALL TREES AND ARE
NEVER FAR FROM
BODIES OF WATER
WHETHER FRESH OR
SALT.

MANGROVES

MANGROVES ARE A TYPE OF TIDAL
WETLAND FOUND ALONG THE COAST
IN WARM PLACES.

MANGROVES

MANGROVES ARE SPECIAL BECAUSE THEY HAVE TREES THAT CAN LIVE IN SALTY WATER WHERE THE OCEAN TIDES COME IN AND OUT.

MANGROVES

ALLIGATORS, FROGS, AND MANY OTHER ANIMALS LIVE IN THESE SWAMPS.

MANGROVES

LEMON SHARKS, WITH THEIR YELLOWISH SKIN, ARE OFTEN FOUND IN MANGROVES-ESPECIALLY WHEN THEY ARE YOUNG-BECAUSE MANGROVES GIVE THEM A SAFE PLACE TO GROW AND HIDE FROM PREDATORS.

MANGROVES

MANGROVE FORESTS ARE COASTAL FORESTS AND CRITICAL HABITATS THAT ACT AS NURSERIES AND PROTECT FROM COASTS FROM EROSION.

NON-TIDAL WETLANDS

NON-TIDAL WETLANDS USUALLY HAVE FRESHWATER FROM RIVERS, LAKES, OR RAIN AND ARE NOT CONNECTED TO THE OCEAN.

CORAL REEFS

CORAL REEFS ARE AN INTEGRAL
PART OF MARINE ECOSYSTEMS.
THEY COVER LESS THAN ONE
PERCENT OF THE OCEAN FLOOR.

CORAL REEFS

CORAL REEFS

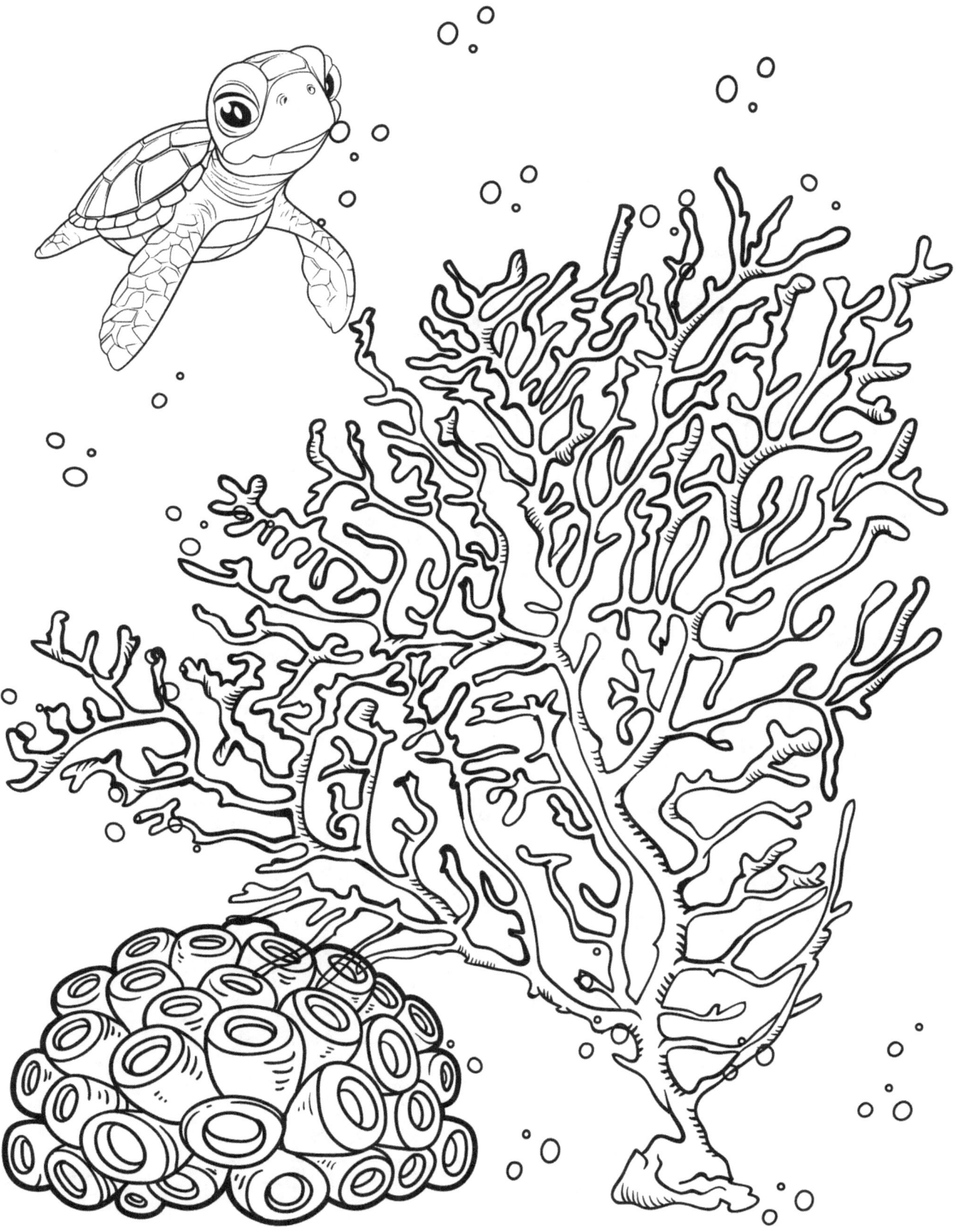

CORAL REEFS

CONCHS ARE SEA ANIMALS WITH BIG
SPIRAL SHELLS, AND THEY OFTEN
LIVE NEAR CORAL REEFS.

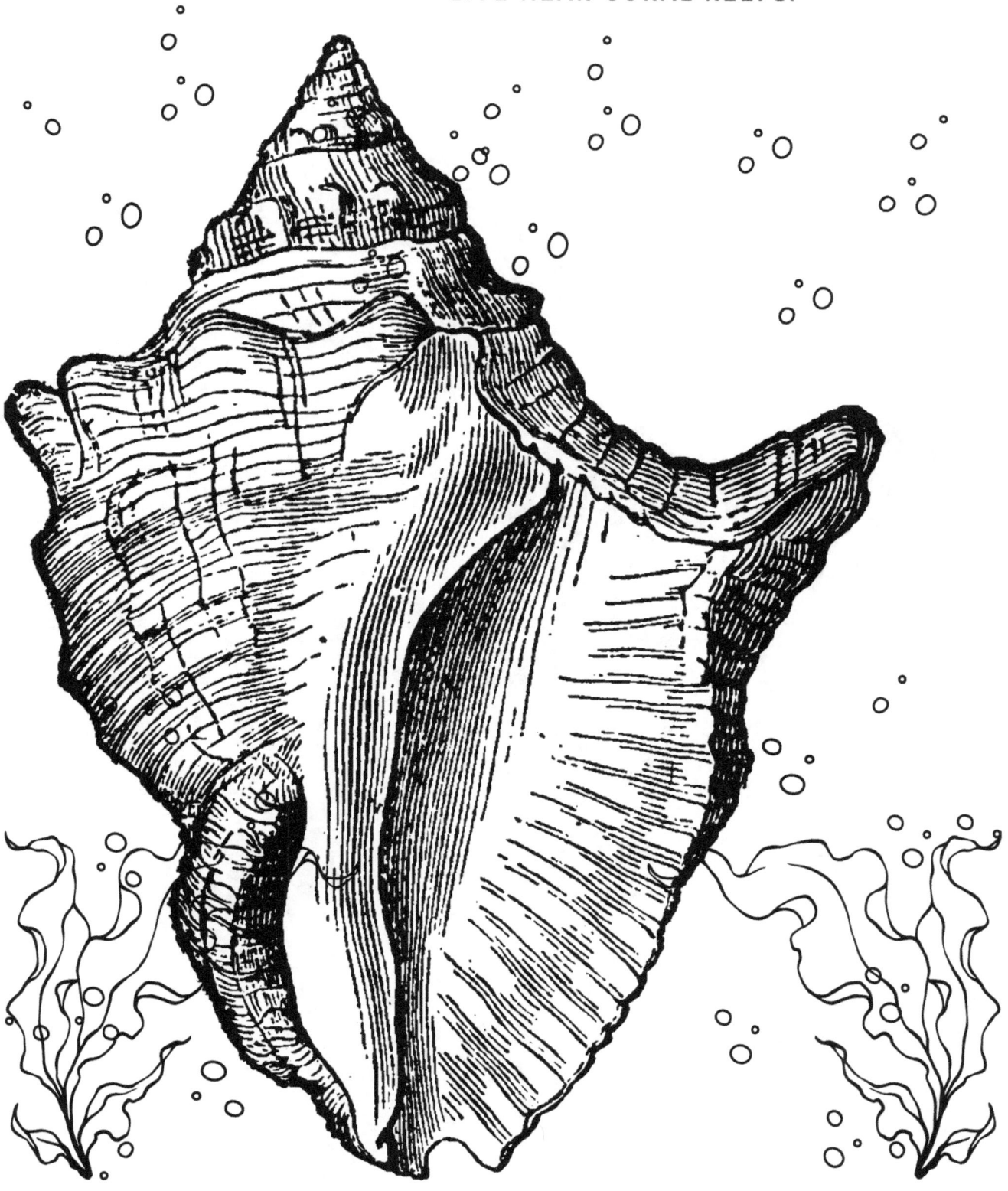

CORAL REEFS

CONCHS HELP KEEP THE REEF CLEAN BY EATING ALGAE AND ARE ALSO FOOD FOR BIGGER ANIMALS, WHICH HELPS THE REEF STAY HEALTHY.

CORAL REEFS

MORAY EELS HUNT FOR FISH IN
SMALL CREVICES ALONG CORAL
REEFS AND SHORELINES

CORAL REEFS

CORAL REEFS

THERE ARE SOME 2,000 SPECIES OF
SEA STAR LIVING IN ALL THE
WORLD'S OCEANS, FROM TROPICAL
HABITATS TO THE COLD SEAFLOOR.

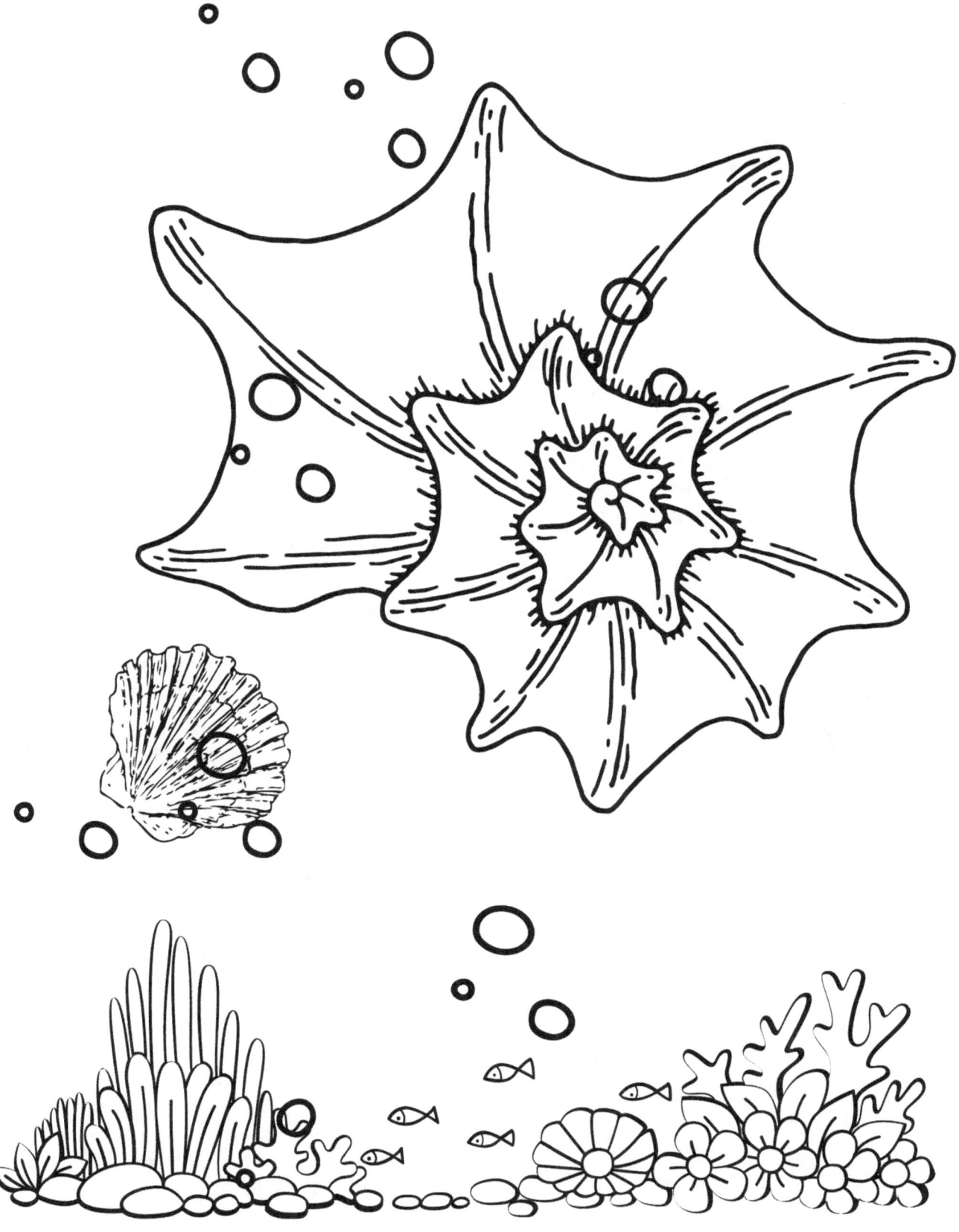

CORAL REEFS

TOURISTS SNORKEL OR DIVE TO SEE
AMAZING UNDERWATER LIFE IN
SOME OF THE WORLD'S MOST
BEAUTIFUL OCEAN PLACES.

BARRIER BEACHES

SOME BEACHES, CALLED BARRIER BEACHES, PROTECT THE MAINLAND FROM THE BATTERING OF OCEAN WAVES.

BARRIER BEACHES

BARRIER BEACHES

BARRIER BEACHES ARE CONSTANTLY
CHANGING, WITH WIND AND WAVES
SHIFTING SAND AND RESHAPING THE
COASTLINE.

BARRIER BEACHES

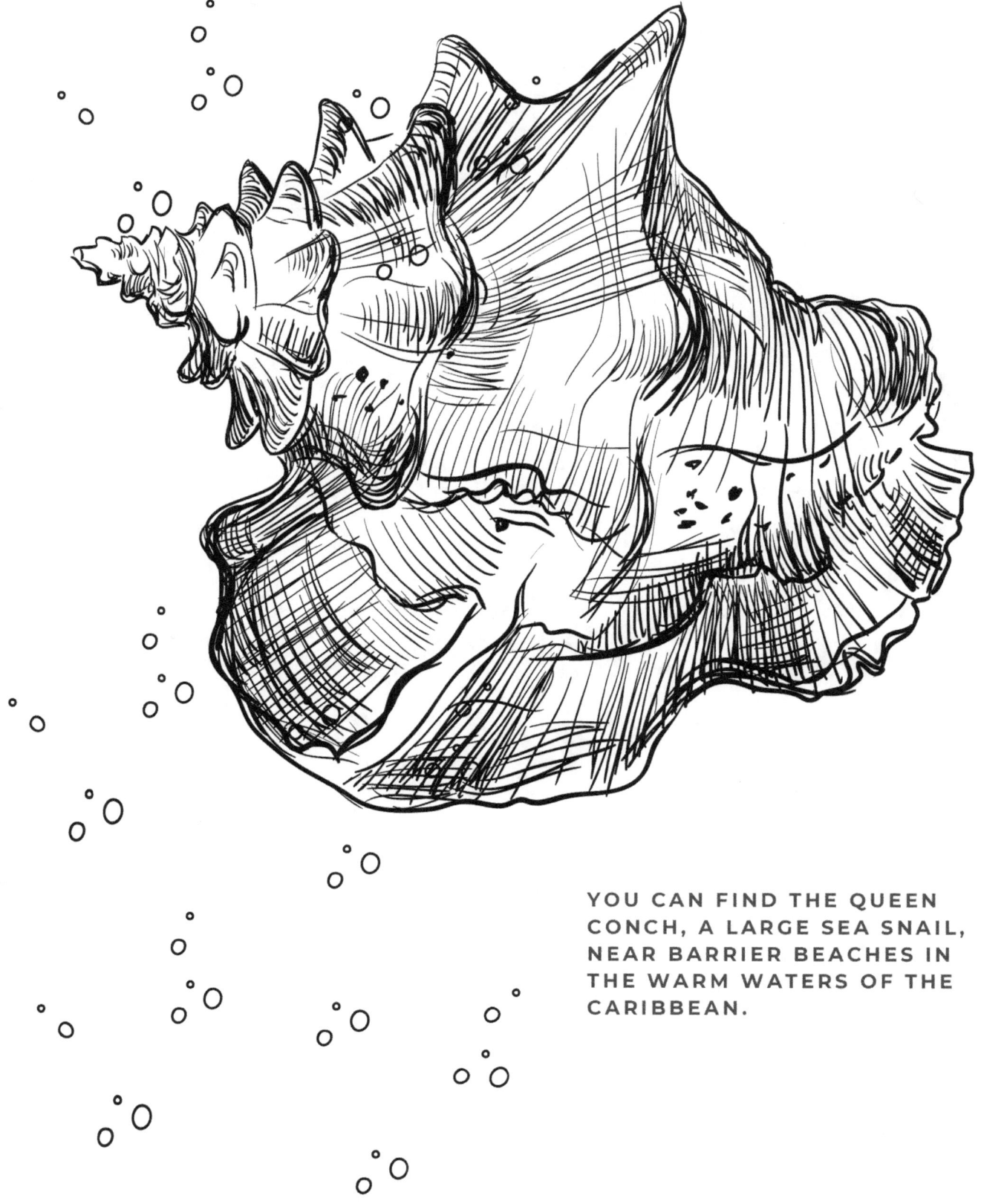

YOU CAN FIND THE QUEEN CONCH, A LARGE SEA SNAIL, NEAR BARRIER BEACHES IN THE WARM WATERS OF THE CARIBBEAN.

BARRIER BEACHES

BARRIER BEACHES ARE IMPORTANT HABITATS FOR A VARIETY OF PLANTS AND ANIMALS, INCLUDING SEABIRDS, FISH, AND SHELLFISH.

BARRIER BEACHES

WHALES CAN SOMETIMES BE SEEN NEAR
BARRIER BEACHES, WHERE THEY MIGHT
COME TO FIND FOOD, TRAVEL, CARE FOR
THEIR BABIES.

BARRIER BEACHES

BLUE WHALES LIVE IN ALL OCEANS EXCEPT THE ARCTIC OCEAN. SOMETIMES, THEY SWIM CLOSE TO THE SHORE, EVEN NEAR BEACHES.

BARRIER BEACHES

JELLYFISH ARE OFTEN FOUND NEAR BARRIER
BEACHES BECAUSE THEY DRIFT WITH OCEAN
CURRENTS AND FOLLOW WARM, SHALLOW
WATERS WHERE THEY CAN FIND FOOD.

BARRIER BEACHES

LOGGERDEAD

THERE ARE SEVEN KINDS OF SEA TURTLES: FLATBACK, GREEN, HAWKSBILL, LEATHERBACK, LOGGERHEAD, KEMP'S RIDLEY, AND OLIVE RIDLEY. SOME NEST ON BARRIER BEACHES.

GREEN TURTLE

BARRIER ISLANDS

BARRIER ISLANDS ARE MADE WHEN WAVES PILE UP SAND NEAR THE SHORE.

BARRIER ISLANDS

BARRIER ISLANDS CAN CHANGE SHAPE,
MOVE, OR EVEN DISAPPEAR BECAUSE
OF WIND, WAVES, AND STORMS.

ARMADILLO

BARRIER ISLANDS

PEARLS CAN BE FOUND IN THE OCEAN NEAR BARRIER ISLANDS. THEY COME FROM PEARL OYSTERS, WHICH LIVE IN SHALLOW, SALTY WATER NEAR THE SHORE.

TIDE POOLS

TIDE POOLS FORM IN ZONES
OF ROCKY SHORELINE

TIDE POOLS

A TIDE POOL IS A SMALL
POOL OF OCEAN WATER
THAT FORMS ON THE
SHORE WHEN THE TIDE
GOES OUT.

TIDE POOLS

SMALL POOLS OF WATER
ARE OFTEN LEFT BEHIND
AMONG THE ROCKS AT
LOW TIDE

TIDE POOLS

TIDE POOLS ARE LITTLE POOLS OF OCEAN WATER WHERE MANY SEA ANIMALS LIVE, LIKE CRABS, STARFISH, SNAILS, AND SMALL FISH.

DEEP SEA

THE DEEP SEA IS THE PART OF THE
OCEAN THAT IS VERY DEEP, FAR
BELOW THE SURFACE WHERE
SUNLIGHT CAN'T REACH.

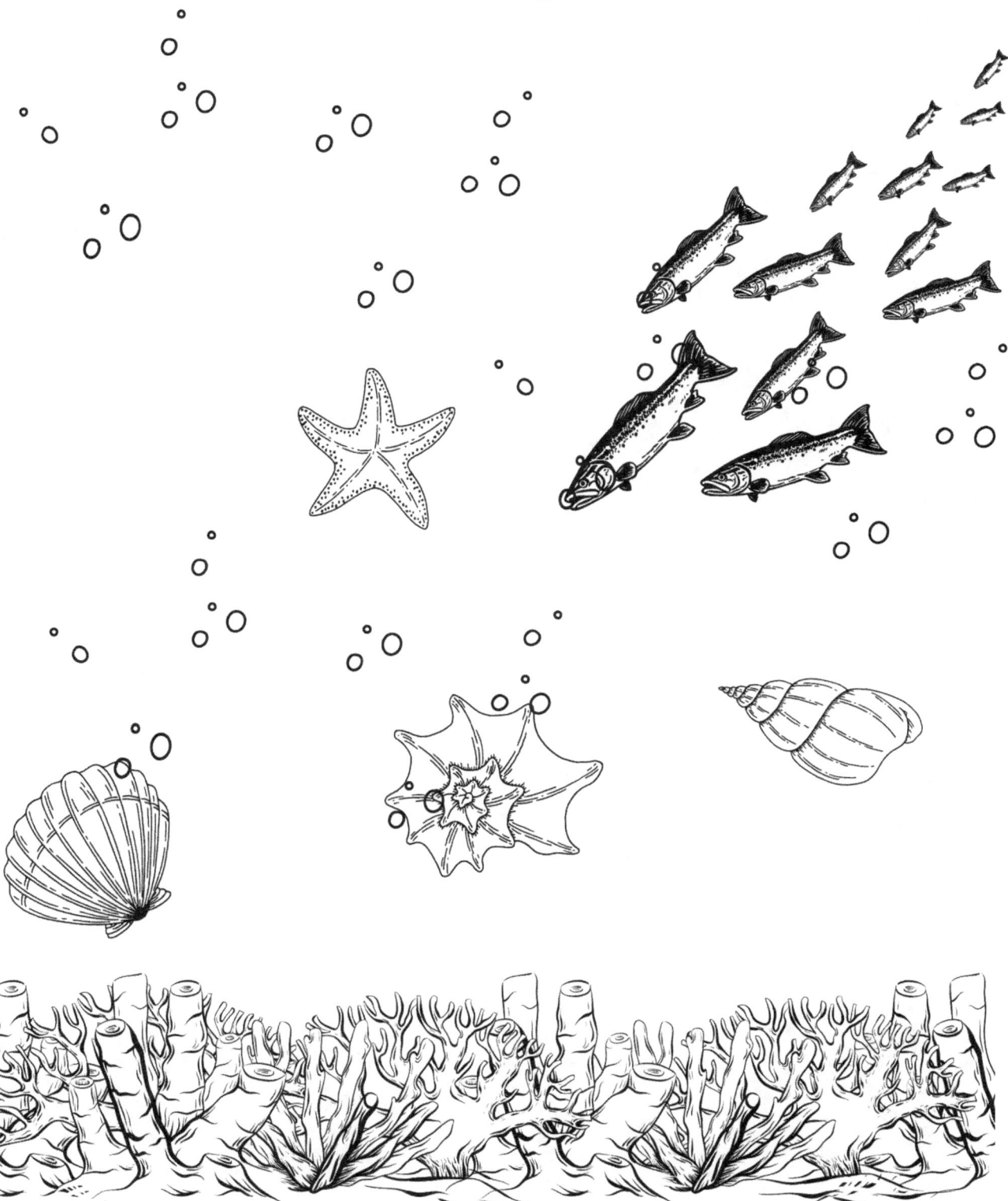

DEEP SEA

DEEP IN THE PACIFIC OCEAN, PALE PURPLE OCTOPUSES WITH GIANT CARTOON EYES ROAM THE SEAFLOOR.

DEEP SEA

DEEP SEA

DEEP SEA

MOST STINGRAYS LIVE IN SHALLOW
OCEAN WATER, BUT SOME, LIKE THE
DEEPWATER STINGRAY, CAN LIVE FAR
DOWN IN THE DEEP SEA.

DEEP SEA

PYGMY SEAHORSES ARE TINY
SEAHORSES THAT LIVE ON CORAL AND
CAN BE FOUND DEEP IN THE OCEAN.
SOME LIVE SO DEEP THAT PEOPLE
HAVE NEVER SEEN THEM ALIVE NEAR
THE TOP!

DEEP SEA

IN THE DEEP SEA, SOME ANIMALS
HELP PLANTS AND ALGAE GROW
BY SPREADING THEIR CELLS-JUST
LIKE BEES HELP FLOWERS.

ANTARCTICA

ANTARCTICA HAS DEEP
OCEAN WATER AROUND IT.

ANTARCTICA

THE OCEAN FLOOR SUPPORTS A RANGE OF SPECIES LIKE SQUID, CRABS, SEA CUCUMBERS, AND SPONGES

ANTARCTICA

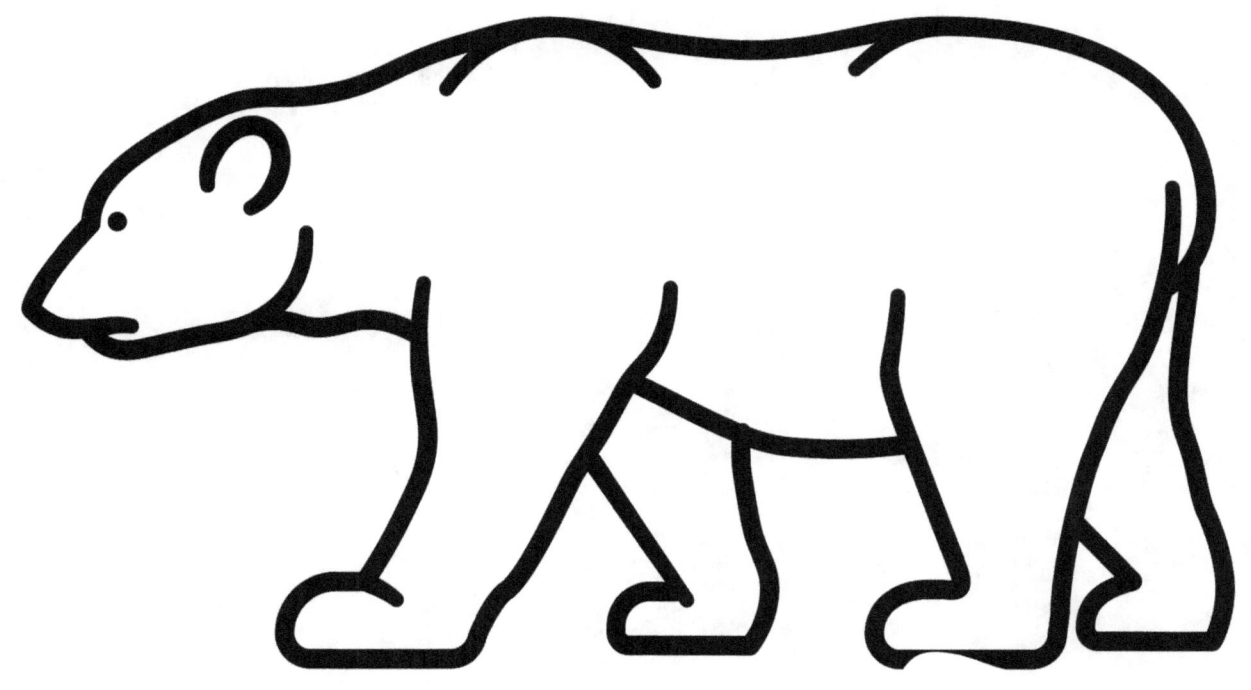

THERE ARE NO POLAR BEARS IN ANTARCTICA! THEIR NATURAL
HABITAT IS THE ARCTIC.

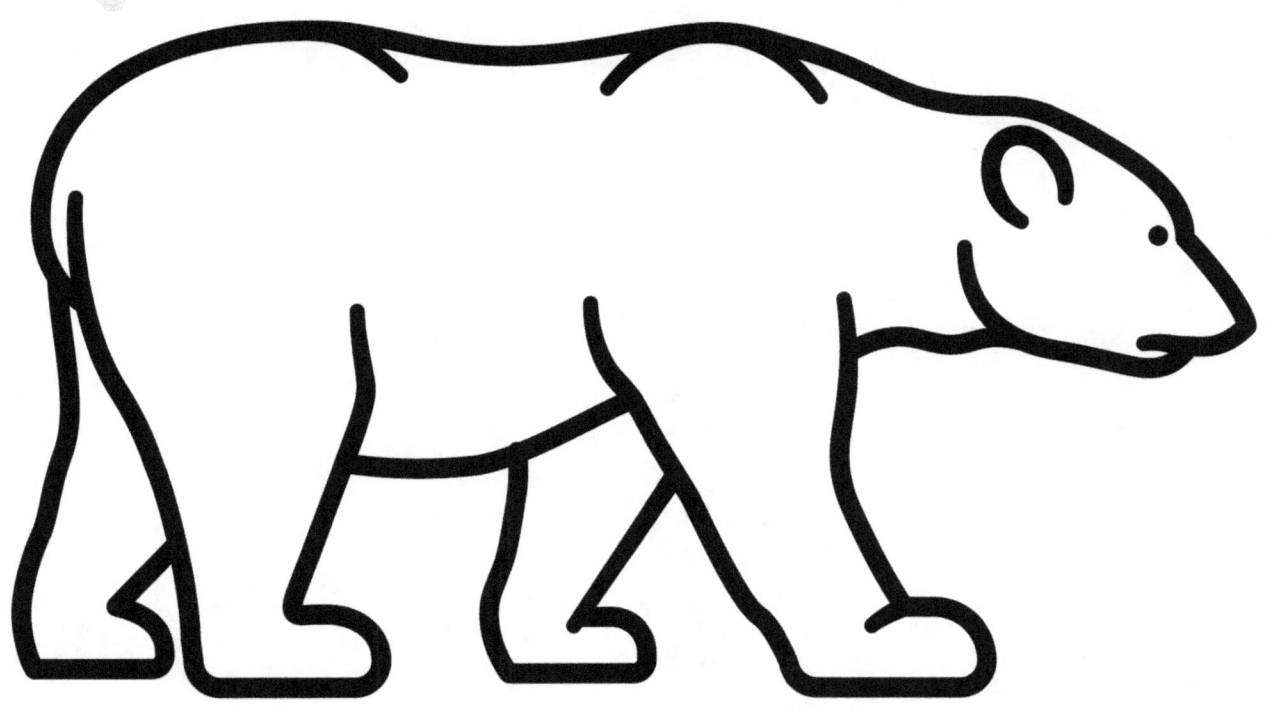

ANTARCTICA

MANY FISH LIVE IN THE COLD OCEAN NEAR
ANTARCTICA. ONE AMAZING FISH IS THE COD
ICEFISH-IT HAS SPECIAL ANTIFREEZE IN ITS
BLOOD SO IT DOESN'T FREEZE!

ANTARCTICA

OCTOPUSES LIVE IN THE COLD SOUTHERN OCEAN AND HAVE SPECIAL PROTEINS IN THEIR BLOOD THAT HELP THEM SURVIVE THE FREEZING TEMPERATURES.

ANTARCTICA

ANTARCTICA'S ECOSYSTEM IS CHANGING DUE TO RISING TEMPERATURES, CAUSING ICE TO MELT AND NEW SPECIES, LIKE KING CRABS, TO MOVE IN.

BIODIVERSITY HOTSPOTS

A WIDE VARIETY OF CICHLID FISH HAVE TURNED EAST AFRICA'S RIFT VALLEY LAKES INTO SOME OF THE RICHEST FRESHWATER ECOSYSTEMS ON EARTH.

BIODIVERSITY HOTSPOT

LAKE VICTORIA IS THE BIGGEST LAKE IN AFRICA AND THE SECOND-BIGGEST FRESHWATER LAKE IN THE WORLD. IT IS SHARED BY THREE COUNTRIES: UGANDA, TANZANIA, AND KENYA.

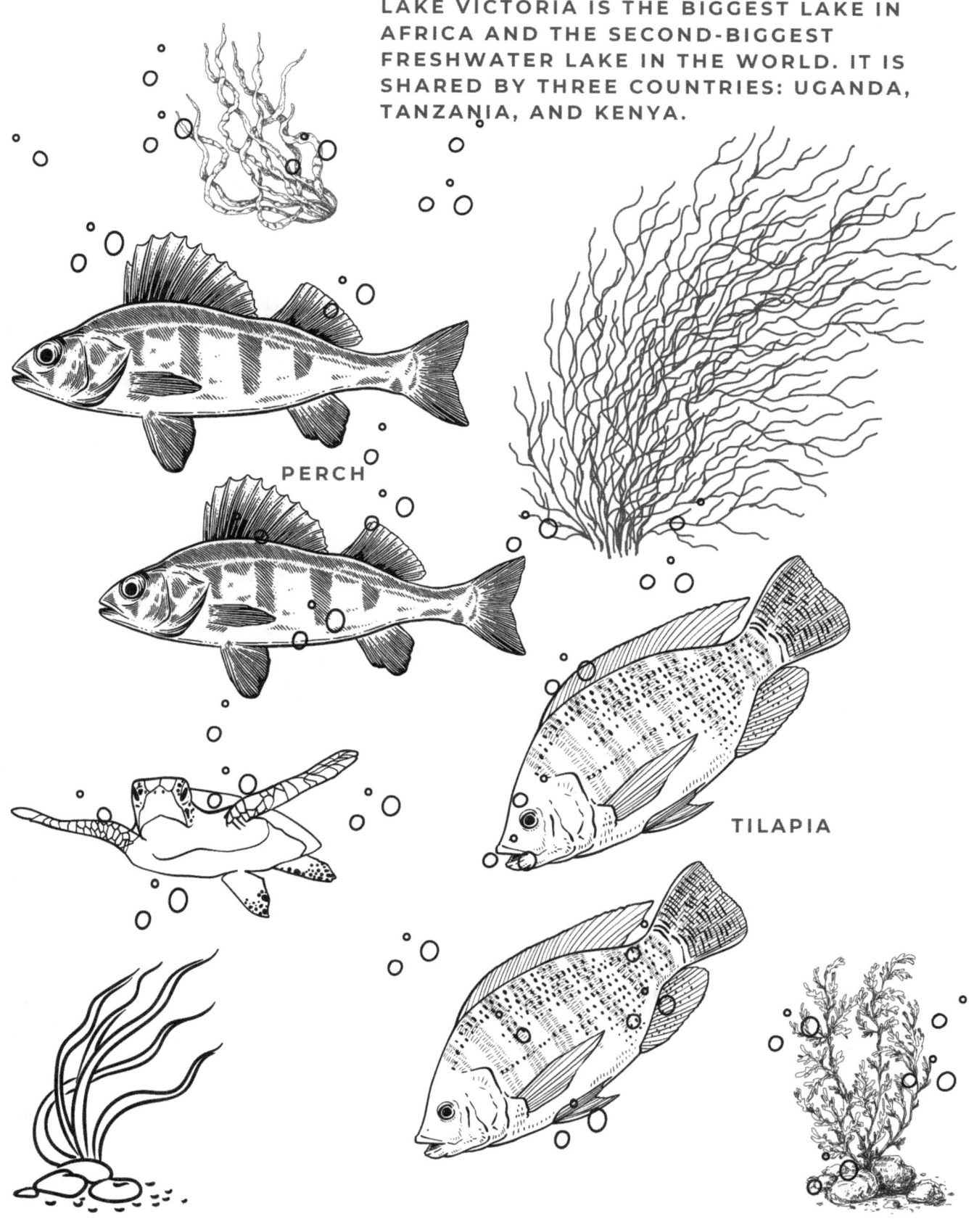

PERCH

TILAPIA

BIODIVERSITY HOTSPOT

THE FIVE GREAT LAKES ARE THE LARGEST GROUP OF FRESHWATER LAKES IN THE WORLD BY SURFACE AREA, LOCATED IN NORTH AMERICA.

WALLEYE

YELLOW PERCH

TROUT

LAKE WHITEFISH

BASS

BIODIVERSITY HOTSPOT

THE CENTRAL INDO-PACIFIC IS AN OCEAN AREA NEAR ASIA AND AUSTRALIA. IT HAS MORE SEA ANIMALS AND CORAL REEFS THAN ANYWHERE ELSE ON EARTH!

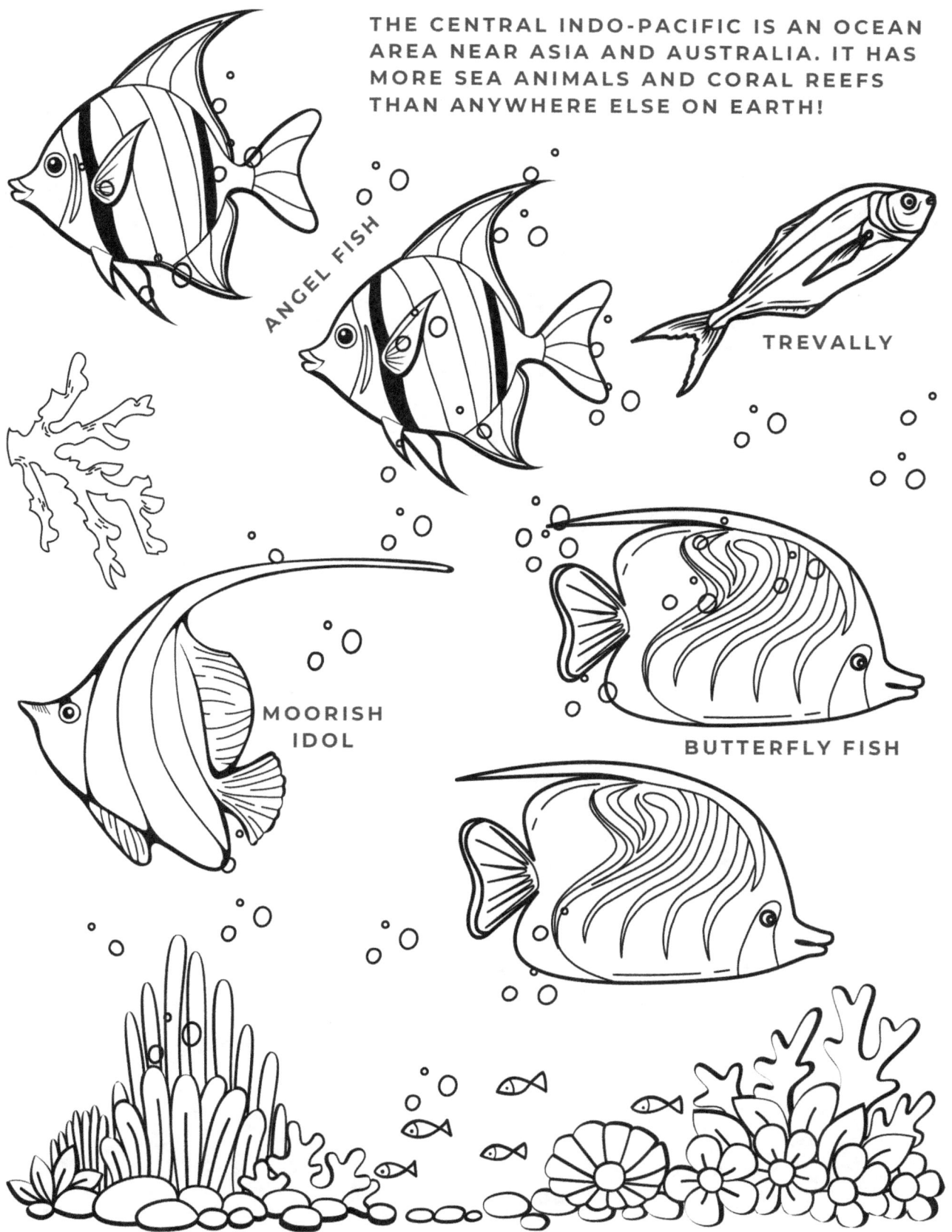

ANGEL FISH

TREVALLY

MOORISH IDOL

BUTTERFLY FISH

BIODIVERSITY HOTSPOT

THE SARGASSO SEA IS A PART OF THE OCEAN WITH LOTS OF FLOATING SEAWEED AND MANY SEA ANIMALS LIVING THERE.

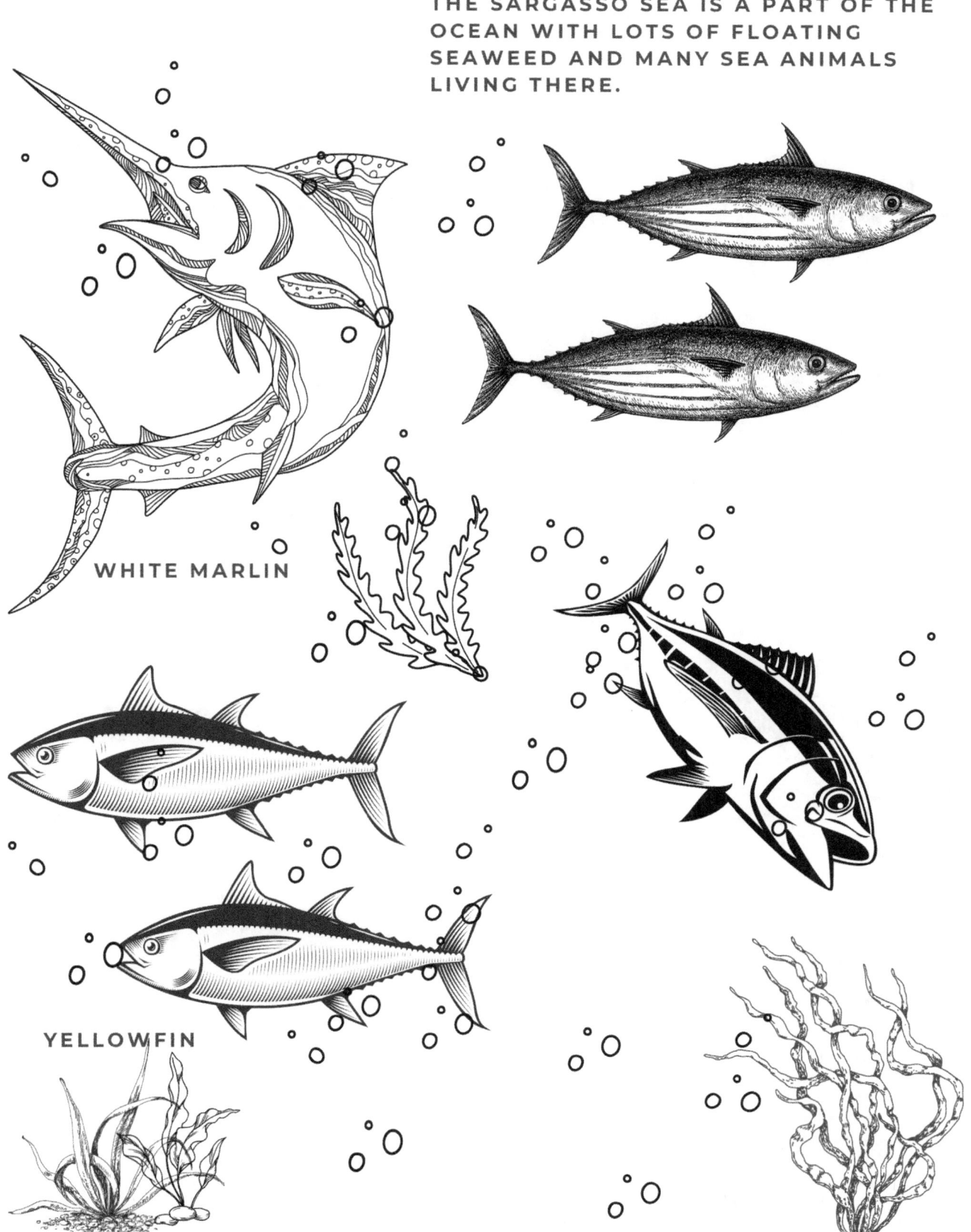

WHITE MARLIN

YELLOWFIN

BIODIVERSITY HOTSPOT

THE BENGUELA CURRENT IS NEAR AFRICA AND HAS LOTS OF OCEAN ANIMALS BECAUSE THE WATER IS FULL OF FOOD AND NUTRIENTS.

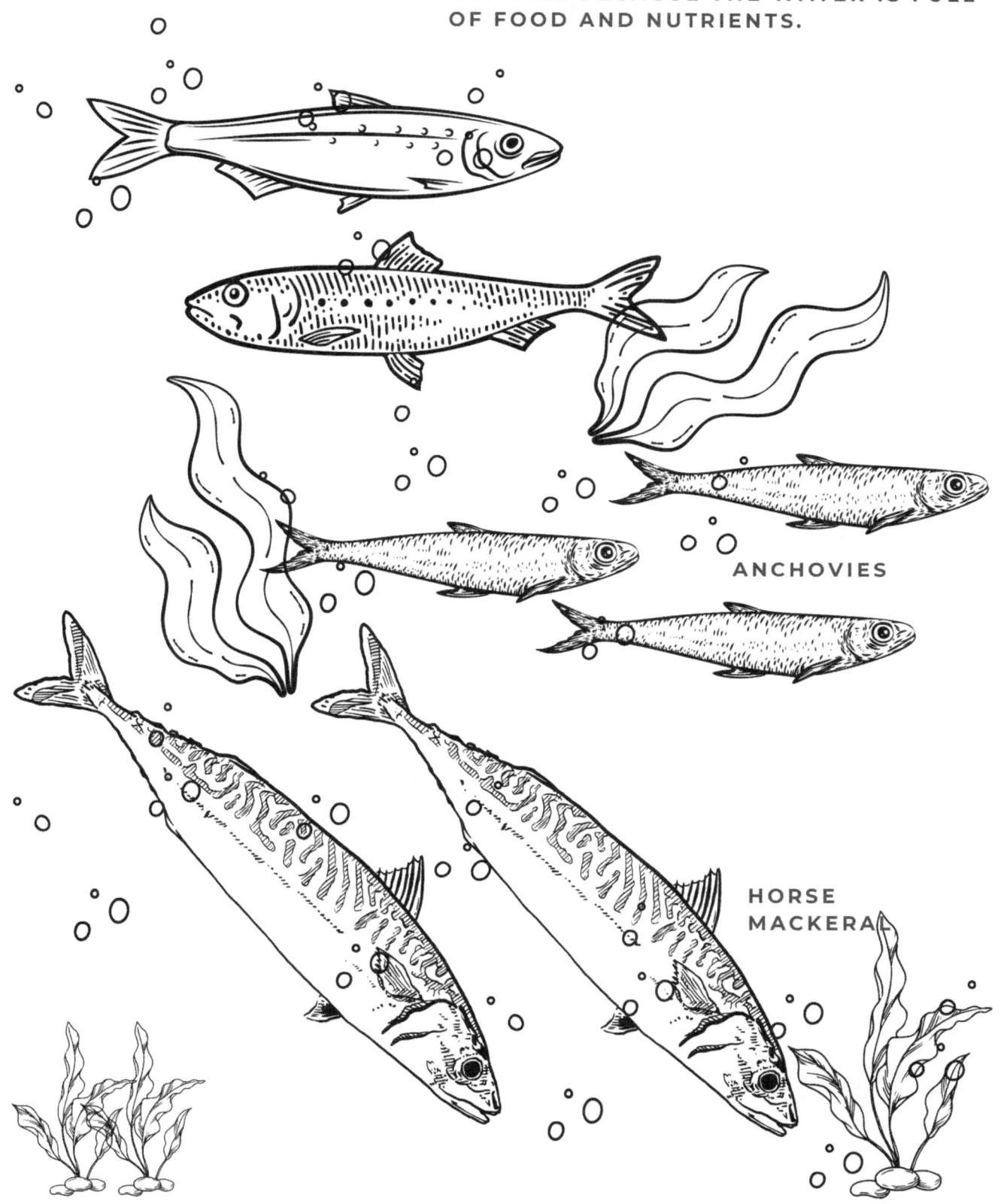

ANCHOVIES

HORSE MACKERAL

CHALLENGES AND RESILIENCE

SOMETIMES OCEAN ANIMALS LOSE
THEIR HOMES BECAUSE OF POLLUTION,
BUILDING NEAR THE SHORE, OR TAKING
TOO MUCH FROM THE OCEAN.

CHALLENGES AND RESILIENCE

CLIMATE CHANGE MAKES THE OCEAN
HOTTER AND HARDER FOR SEA ANIMALS
TO LIVE IN.

SEA LION

CHALLENGES AND RESILIENCE

INVASIVE SPECIES ARE ANIMALS OR PLANTS THAT COME FROM SOMEWHERE ELSE AND CAN CAUSE TROUBLE FOR THE ONES ALREADY LIVING THERE.

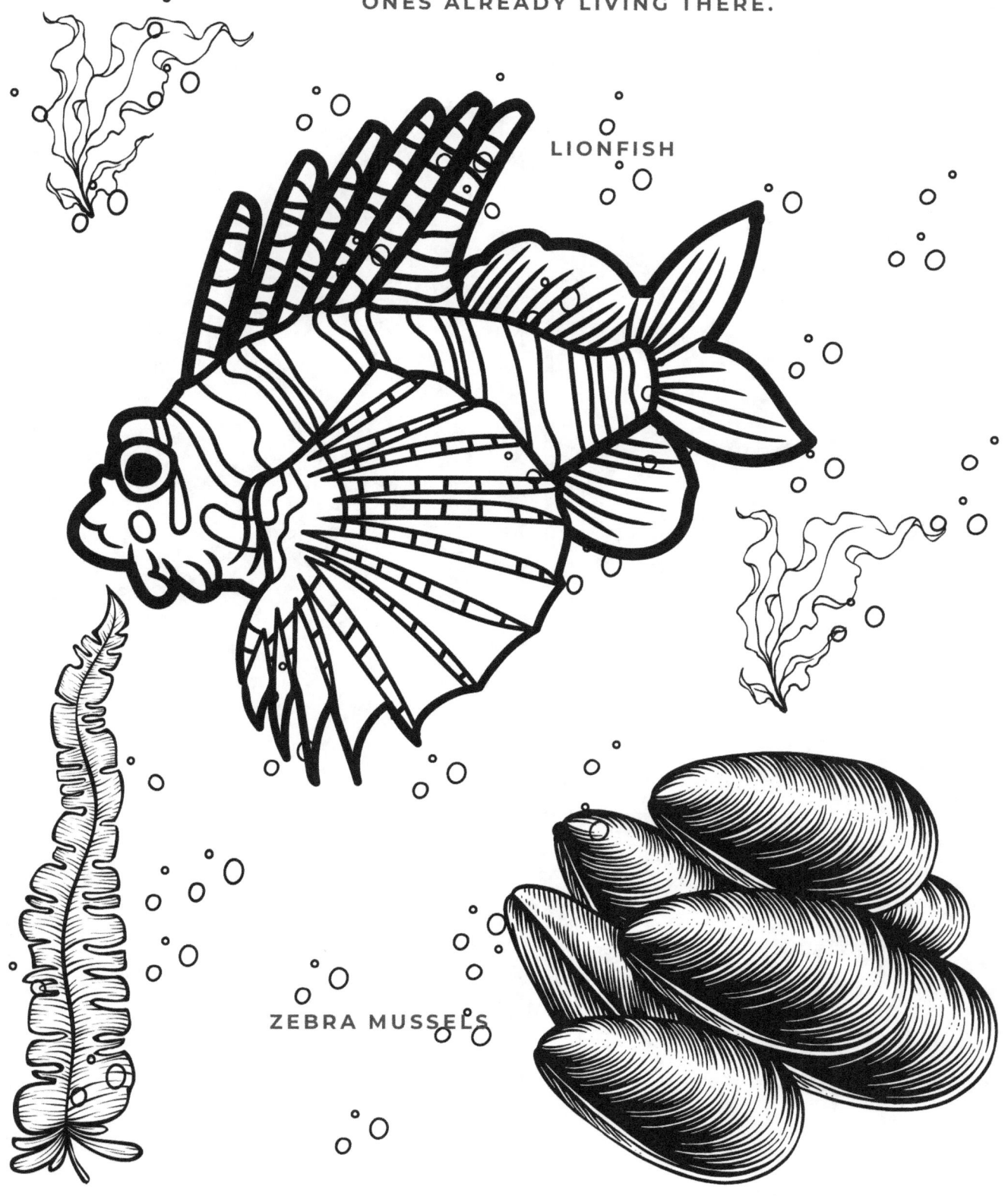

LIONFISH

ZEBRA MUSSELS

CHALLENGES AND RESILIENCE

CATCHING TOO MANY FISH AND DIRTY
WATER CAN HURT OCEAN ANIMALS AND
THEIR HOMES.

MARINE ECONOMICS

MARINE ECONOMICS IS THE STUDY OF HOW
HUMANS USE AND MANAGE OCEAN AND
COASTAL RESOURCES FOR ECONOMIC BENEFIT.

MARINE ECONOMICS

MARINE ECONOMICS FOCUSES ON THE VALUE
OF GOODS AND SERVICES PROVIDED BY THE
MARINE ENVIRONMENT AND HOW TO
BALANCE ECONOMIC GROWTH WITH
SUSTAINABILITY.

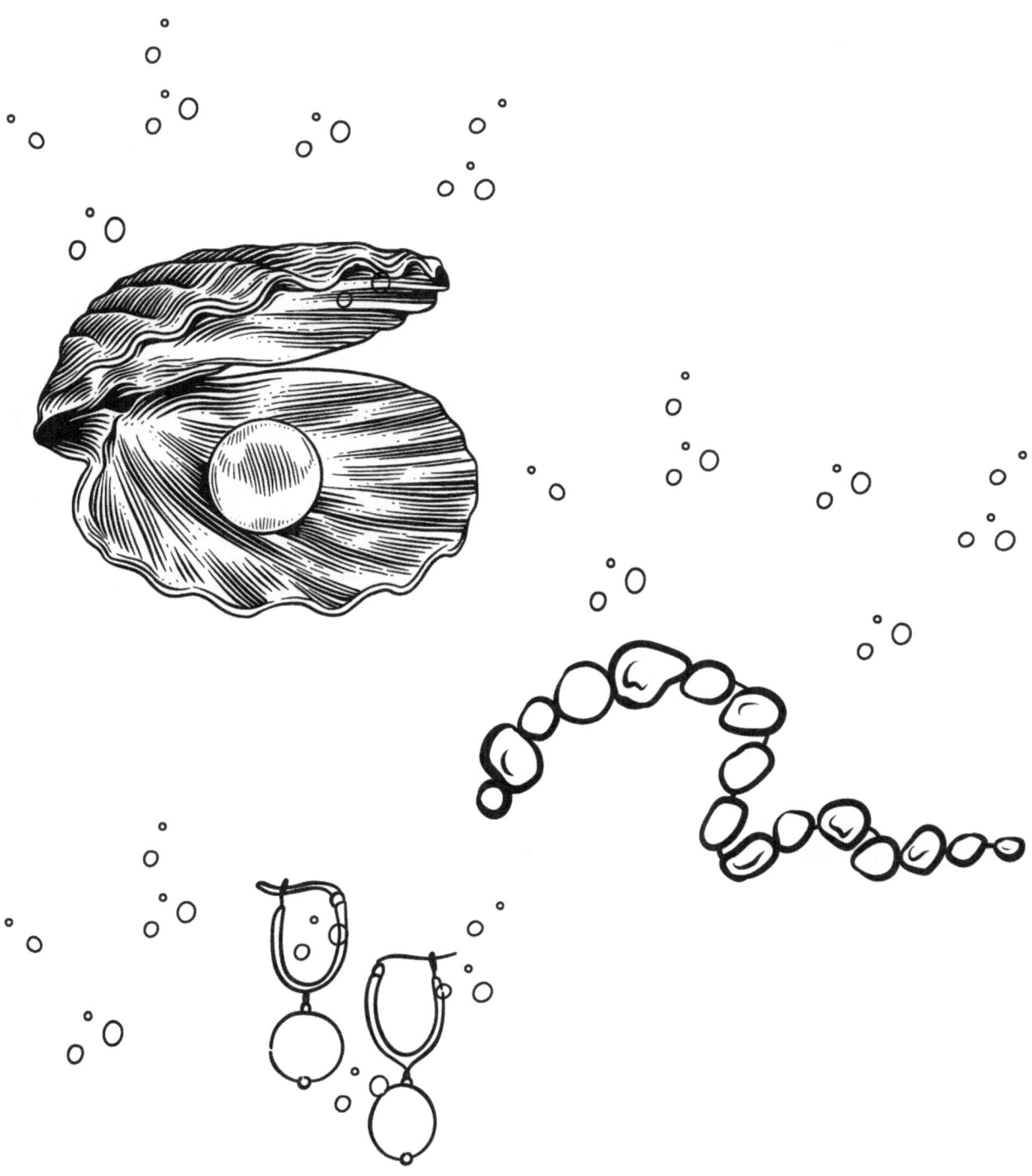

CONSERVATION EFFORTS

MARINE PROTECTED AREAS (MPA) ARE OCEAN PLACES WHERE ANIMALS AND PLANTS ARE KEPT SAFE.

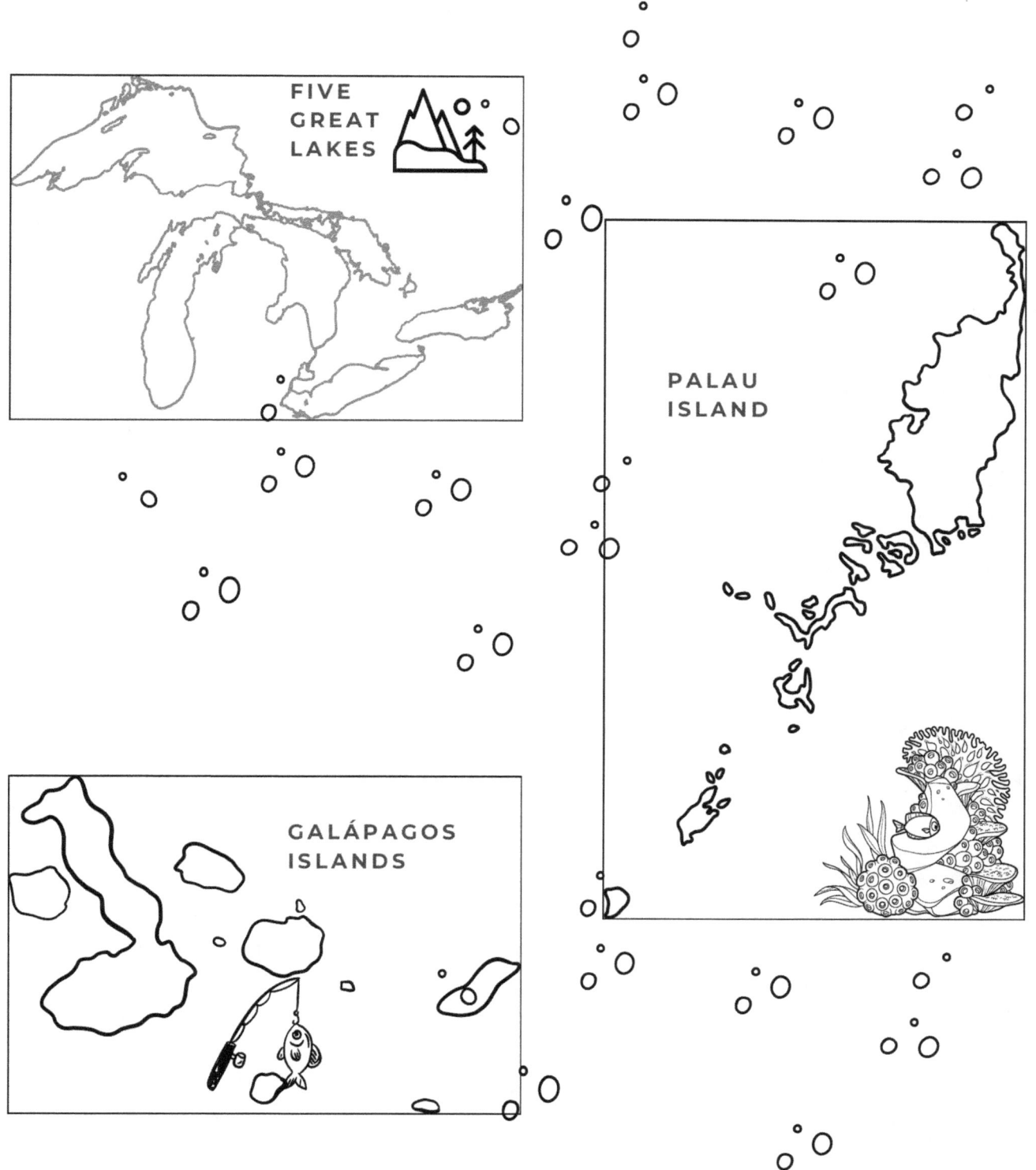

FIVE GREAT LAKES

PALAU ISLAND

GALÁPAGOS ISLANDS

CONSERVATION EFFORTS

SUSTAINABLE FISHING MEANS CATCHING
ONLY SOME FISH SO THE OCEAN STAYS
HEALTHY AND FULL OF FISH.

CONSERVATION EFFORTS

SCIENTISTS WATCH AND STUDY THE
OCEAN TO HELP PROTECT SEA ANIMALS
AND PLANTS.

CONSERVATION EFFORTS

RESEARCH PLAYS A CRUCIAL ROLE IN
UNDERSTANDING WATER ECOSYSTEMS,
ENVIRONMENTAL CHALLENGES, AND HOW TO
PROTECT THESE VITAL RESOURCES

WATER, WATER, EVERYWHERE!

ABOUT 71% OF EARTH'S SURFACE IS WATER, WITH 97% BEING SALTY OCEAN WATER. ONLY 3% IS FRESHWATER, MOST OF WHICH IS FROZEN IN GLACIERS OR UNDERGROUND.

NOTES